WILLIAM H

GREAT ASTRONOMERS

ROBERT STAWELL BALL

GIFT *Certificate*

TO:

FROM:

DATE:

WILLIAM HERSCHEL

GREAT ASTRONOMERS

BY

ROBERT STAWELL BALL

D.Sc, LL.D., F.R.S.

(July 1, 1840 – November 25, 1913)

LOWNDEAN PROFESSOR OF ASTRONOMY AND GEOMETRY
IN THE UNIVERSITY OF CAMBRIDGE

DIAMOND™ BOOKS

www.diamondbooks.ca

TORONTO, CANADA – 2017

DIAMOND BOOKS - CANADA

DIAMOND BOOKS

Toronto, ON, CANADA

http://www.diamondbooks.ca

BIBLIOGRAPHIC INFORMATION

'WILLIAM HERSCHEL'
Is an extract or a chapter from the book
"GREAT ASTRONOMERS" by Robert Stawell Ball.

FIRST PUBLISHED IN: 1896.

PUBLISHED IN CANADA

Published in Canada by DIAMOND BOOKS - CANADA, an imprint of DIAMOND PUBLISHERS - http://www.diamondpublishers.com

FIRST EDITION: MARCH, 2017.

PAPERBACK EDITION : ISBN: 978-1-988357-55-3

PRINTED IN CANADA

PREFACE
TO THE ORIGINAL EDITION

It has been my object in these pages to present the life of each astronomer in such detail as to enable the reader to realise in some degree the man's character and surroundings; and I have endeavoured to indicate as clearly as circumstances would permit the main features of the discoveries by which he has become known.

There are many types of astronomers—from the stargazer who merely watches the heavens, to the abstract mathematician who merely works at his desk; it has, consequently, been necessary in the case of some lives to adopt a very different treatment from that which seemed suitable for others.

ROBERT STAWELL BALL.

The Observatory, Cambridge.
October, 1895.

NOTE: The full-length of 'Preface' can be found in its original edition of the book "GREAT ASTRONOMERS".

INTRODUCTION

Of all the natural sciences there is not one which offers such sublime objects to the attention of the inquirer as does the science of astronomy. From the earliest ages the study of the stars has exercised the same fascination as it possesses at the present day. Among the most primitive peoples, the movements of the sun, the moon, and the stars commanded attention from their supposed influence on human affairs. The practical utilities of astronomy were also obvious in primeval times. Maxims of extreme antiquity show how the avocations of the husbandman are to be guided by the movements of the heavenly bodies. The positions of the stars indicated the time to plough, and the time to sow. To the mariner who was seeking a way across the trackless ocean, the heavenly bodies .offered the only reliable marks by which his path could be guided. There was, accordingly, a stimulus both from intellectual curiosity and from practical necessity to follow the movements of the stars. Thus began a search for the causes of the ever-varying phenomena which the heavens display.

- ROBERT STAWELL BALL

Many of the earliest discoveries are indeed prehistoric. The great diurnal movement of the heavens, and the annual revolution of the sun, seem to have been known in times far more ancient than those to which any human monuments can be referred. The acuteness of the early observers enabled them to single out the more important of the wanderers which we now call planets. They saw that the star-like objects, Jupiter, Saturn, and Mars, with the more conspicuous Venus, constituted a class of bodies wholly distinct from the fixed stars among which their movements lay, and to which they bear such a superficial resemblance. But the penetration of the early astronomers went even further, for they recognized that Mercury also belongs to the same group, though this particular object is seen so rarely. It would seem that eclipses and other phenomena were observed at Babylon from a very remote period, while the most ancient records of celestial observations that we possess are to be found in the Chinese annals. The study of astronomy, in the sense in which we understand the word, may be said to have commenced under the reign of the Ptolemies at Alexandria. The most famous name in the science of this period is that of Hipparchus, who lived and worked at Rhodes about the year 160 B.C. It was his splendid investigations that first wrought the observed facts into a coherent branch of knowledge. He recognized the primary obligation which lies on the student of the heavens to compile as complete an inventory as possible of the objects which are there to be found. Hipparchus accordingly commenced by undertaking, on a small scale, a task exactly similar to that on which modern astronomers, with all available appliances of meridian circles, and photographic telescopes, are constantly engaged at the present day. He compiled a catalogue of the principal fixed stars, which is of special value to astronomers, as being the earliest work of its kind which has been handed down. He also studied the movements of the sun and the moon, and framed theories to account for the incessant changes which he saw in progress. He found a much more difficult problem in his attempt to interpret satisfactorily the complicated movements of the planets. With the view of constructing a theory which should give some coherent account of the subject, he made many observations of the places of these wandering stars. How great were the advances which Hipparchus accomplished may be appreciated if we reflect that, as a preliminary task to his more purely astronomical labours, he had to invent that branch of mathematical science by which alone the problems he proposed could be solved. It was for this purpose that he devised the indispensable method of calculation which we now know so well as trigonometry. Without the aid rendered by this beautiful art it would have been impossible for any n ally important advance in astronomical calculation to have been effected.

But the discovery which shows, beyond all others, that Hipparchus possessed one of the master-minds of all time was the detection of that remarkable celestial movement known as the precession of the equinoxes. The inquiry which conducted to this discovery involved a most profound investigation, especially when it is remembered that in the days of Hipparchus the means of observation of the heavenly bodies were only of the rudest description, and the available observations of earlier dates were extremely scanty. We can but look with astonishment on the genius of the man who, in spite of such difficulties, was able to detect such a phenomenon as the precession, and to exhibit its actual magnitude. I shall endeavour to explain the nature of this singular celestial movement, for it may be said to offer the first instance in the history of science in which we find that combination of accurate observation with skilful interpretation, of which, in the subsequent development of astronomy, we have so many splendid examples.

The word equinox implies the condition that the night is equal to the day. To a resident on the equator the night is no doubt equal to the day at all times in the year, but to one who lives on any other part of the earth, in either hemisphere, the night and the day are not generally equal. There is, however, one occasion in spring, and another in autumn, on which the day and the night are each twelve hours at all places on the earth. When the night and day are equal in spring, the point which the sun occupies on. the heavens is termed the vernal equinox. There is similarly another point in which the sun is situated at the time of the autumnal equinox. In any investigation of the celestial movements the positions of these two equinoxes on the heavens are of primary importance, and Hipparchus, with the instinct of genius, perceived their significance, and commenced to study them. It will be understood that we can always define the position of a point on the sky with reference to the surrounding stars. No doubt we do not see the stars near the sun when the sun is shining, but they are there nevertheless. The ingenuity of Hipparchus enabled him to determine the positions of each of the two equinoxes relatively to the stars which lie in its immediate vicinity. After examination of the celestial places of these points at different periods, he was led to the conclusion that each equinox was moving relatively to the stars, though that movement was so slow that twenty five thousand years would necessarily elapse before a complete circuit of the heavens was accomplished. Hipparchus traced out this phenomenon, and established it on an impregnable basis, so that all astronomers have ever since recognised the precession of the equinoxes) as one of the fundamental facts of astronomy. Not until nearly two thousand years after Hipparchus had made this splendid discovery was the explanation of its cause given by Newton.

From the days of Hipparchus down to the present hour the science of astronomy has steadily grown. One great observer after another has appeared from time to time, to reveal some new phenomenon with regard to the celestial bodies or their movements, while from time to time one commanding intellect after another has arisen to explain the true import of the facts of observations. The history of astronomy thus becomes inseparable from the history of the great men to whose labours its development is due. In the ensuing chapters we have endeavoured to sketch the lives and the work of the great philosophers, by whose labours the science of astronomy has been created.

We shall commence with Ptolemy, who, after the foundations of the science had been laid by Hipparchus, gave to astronomy the form in which it was taught throughout the Middle Ages. We shall next see the mighty revolution in our conceptions of the universe which are associated with the name of Copernicus. We then pass to those periods illumined by the genius of Galileo and Newton, and afterwards we shall trace the careers of other more recent discoverers, by whose industry and genius the boundaries of human knowledge have been so greatly extended. Our history will be brought down late enough to include some of the illustrious astronomers who laboured in the generation which has just passed away.

- ROBERT STAWELL BALL

TABLE OF CONTENTS

CHAPTERS

1. **Chapter # 10***
   William Herschel 1

2. A Brief Biographical Sketch 35
   - *Complimentary Material*

ABOUT THE AUTHOR 75

**An original Chapter from the book:*
"GREAT ASTRONOMERS", by Robert Stawell Ball.

CHAPTER

10

GREAT ASTRONOMERS

William Herschel

(November 15, 1738 – August 25, 1822)

William Herschel

(November 15, 1738 – August 25, 1822)

CHAPTER 10

GREAT ASTRONOMERS: WILLIAM HERSCHEL

WILLIAM HERSCHEL, one of the greatest astronomers that has ever lived, was born at Hanover, on the 15th November, 1738. His father, Isaac Herschel, was a man evidently of considerable ability, whose life was devoted to the study and practice of music, by which he earned a somewhat precarious maintenance. He had but few worldly goods to leave to his children, but he more than compensated for this by bequeathing to them a splendid inheritance of genius. Touches of genius were, indeed, liberally scattered among the members of Isaac's large family, and in the case of his fourth child, William, and of a sister several years younger, it was united with that determined perseverance and rigid adherence to principle which enabled genius to fulfill its perfect work.

A faithful chronicler has given us an interesting account of the way in which Isaac Herschel educated his sons; the narrative is taken from the recollections of one who, at the time we are speaking of, was an unnoticed little girl five or six years old. She writes:—

"My brothers were often introduced as solo performers and assistants in the orchestra at the Court, and I remember that I was frequently prevented from going to sleep by the lively criticisms on music on coming from a concert. Often I would keep myself awake that I might listen to their animating remarks, for it made me so happy to see them so happy. But generally their conversation would branch out on philosophical subjects, when my brother

William and my father often argued with such warmth that my mother's interference became necessary, when the names—Euler, Leibnitz, and Newton—sounded rather too loud for the repose of her little ones, who had to be at school by seven in the morning." The child whose reminiscences are here given became afterwards the famous Caroline Herschel. The narrative of her life, by Mrs. John Herschel, is a most interesting book, not only for the account it contains of the remarkable woman herself, but also because it provides the best picture we have of the great astronomer to whom Caroline devoted her life.

This modest family circle was, in a measure, dispersed at the outbreak of the Seven Years' War in 1756. The French proceeded to invade Hanover, which, it will be remembered, belonged at this time to the British dominions. Young William

Herschel had already obtained the position of a regular performer in the regimental band of the Hanoverian Guards, and it was his fortune to obtain some experience of actual warfare in the disastrous battle of Hastenbeck. He was not wounded, but he had to spend the night after the battle in a ditch, and his meditations on the occasion convinced him that soldiering was not the profession exactly adapted to his tastes. We need not attempt to conceal the fact that he left his regiment by the very simple but somewhat risky process of desertion. He had, it would seem, to adopt disguises to effect his escape. At all events, by some means he succeeded in eluding detection and reached England in safety. It is interesting to have learned on good authority that many years after this offence was committed it was solemnly forgiven. When Herschel had become the famous astronomer, and as such visited King George at Windsor, the King at their first meeting handed to him his pardon for deserting from the army, written out in due form by his Majesty himself.

It seems that the young musician must have had some difficulty in providing for his maintenance during the first few years of his abode in England. It was not until he had reached the age of twenty-two that he succeeded in obtaining any regular appointment. He was then made Instructor of Music to the Durham Militia. Shortly afterwards, his talents being more widely recognised, he was appointed as organist at the parish church at Halifax, and his prospects in life now being fairly favourable, and the Seven Years' War being over, he ventured to pay a visit to Hanover to see his father. We can imagine the delight with which old Isaac Herschel welcomed his promising son, as well as his parental pride when a concert was given at which some of William's compositions were performed. If the father was so intensely gratified on this occasion, what would his feelings have been could he have lived to witness his son's future career? But this pleasure was not to be his, for he died many years before William became an astronomer.

In 1766, about a couple of years after his return to England from this visit to his old home, we find that Herschel had received a further promotion to be organist in the Octagon Chapel, at Bath. Bath was then, as now, a highly fashionable resort, and many notable personages patronised the rising musician. Herschel had other points in his favour besides his professional skill; his appearance was good, his address was prepossessing, and even his nationality was a distinct advantage, inasmuch as he was a Hanoverian in the reign of King George the Third. On Sundays he played the organ, to the great delight of the congregation, and on week-days he was occupied by giving lessons to private pupils, and in preparation for public performances. He thus came to be busily employed, and seems to have been in the enjoyment of comfortable means.

From his earliest youth Herschel had been endowed with that invaluable characteristic, an eager curiosity for knowledge. He was naturally desirous of perfecting himself in the theory of music, and thus he was led to study mathematics. When he had once tasted the charms of mathematics, he saw vast regions of knowledge unfolded before him, and in this way he was induced to direct his attention to astronomy. More and more this pursuit seems to have engrossed his attention, until at last it had become an absorbing passion. Herschel was, however, still obliged, by the exigency of procuring a livelihood, to give up the best part of his time to his profession as a musician; but his heart was eagerly fixed on another science, and every spare moment was steadily devoted to astronomy. For many years, however, he continued to labour at his original calling, nor was it until he had attained middle age and become the most celebrated astronomer of the time, that he was enabled to concentrate his attention exclusively on his favourite pursuit.

It was with quite a small telescope which had been lent him by a friend that Herschel commenced his career as an observer.

19, New King Street, Bath, where Herschel lived.

However, he speedily discovered that to see all he wanted to see, a telescope of far greater power would be necessary, and he determined to obtain this more powerful instrument by actually making it with his own hands. At first it may seem scarcely likely that one whose occupation had previously been the study and practice of music should meet with success in so technical an operation as the construction of a telescope. It may, however, be mentioned that the kind of instrument which Herschel designed to construct was formed on a very different principle from the refracting telescopes with which we are ordinarily familiar. His telescope was to be what is termed a reflector.1 In this type of instrument the optical power is obtained by the use of a mirror at the bottom of the tube, and the astronomer looks down through the tube towards his mirror, and views the reflection of the stars with its aid. Its efficiency as a telescope depends entirely on the accuracy with which the requisite form has been imparted to the mirror. The surface has to be hollowed out a little, and this has to be done so truly that the slightest deviation from good workmanship in this essential particular would be fatal to efficient performance of the telescope.

The mirror that Herschel employed was composed of a mixture of two parts of copper to one of tin; the alloy thus obtained is an intensely hard material, very difficult to cast into the proper shape, and very difficult to work afterwards.

William Herschel.

It possesses, however, when polished, a lustre hardly inferior to that of silver itself. Herschel has recorded hardly any particulars as to the actual process by which he cast and figured his reflectors. We are, however, told that in later years, after his telescopes had become famous, he made a considerable sum of money by the manufacture and sale of great instruments. Perhaps this may be the reason why he never found it expedient to publish any very explicit details as to the means by which his remarkable successes were obtained.

Caroline Herschel.

Since Herschel's time many other astronomers, notably the late Earl of Rosse, have experimented in the same direction, and succeeded in making telescopes certainly far greater, and probably more perfect, than any which Herschel appears to have constructed. The details of these later methods are now well known, and have been extensively practised. Many amateurs have thus been able to make telescopes by following the instructions so clearly laid down by Lord Rosse and the other authorities. Indeed, it would seem that anyone who has a little mechanical skill and a good deal of patience ought now to experience no great difficulty in constructing a telescope quite as powerful as that which first brought Herschel into fame. I should, however, mention that in these modern days the material generally used for the mirror is of a more tractable description than the metallic substance which was employed by Herschel and by Lord Rosse. A reflecting telescope of the present day would not be fitted with a mirror composed of that alloy known as speculum metal, whose composition I have already mentioned. It has been found more advantageous to employ a glass mirror carefully figured and polished, just as a metallic mirror would have been, and then to impart to the polished glass surface a fine coating of silver laid down by a chemical process. The silver-on-glass mirrors are so much lighter and so much easier to construct that the more old-fashioned metallic mirrors may be said to have fallen into almost total disuse. In one respect, however, the metallic mirror may still claim the advantage that, with reasonable care, its surface will last bright and untarnished for a much longer period than can the silver film on the glass. However, the operation of re-silvering a glass has now become such a simple one that the advantage this indicates is not relatively so great as might at first be supposed.

- ROBERT STAWELL BALL

Street view, Herschel House, Slough.

Some years elapsed after Herschel's attention had been first directed to astronomy, before he reaped the reward of his exertions in the possession of a telescope which would adequately reveal some of the glories of the heavens. It was in 1774, when the astronomer was thirty-six years old, that he obtained his first glimpse of the stars with an instrument of his own construction. Night after night, as soon as his musical labours were ended, his telescopes were brought out, sometimes into the small back garden of his house at Bath, and sometimes into the street in front of his hall-door. It was characteristic of him that he was always endeavouring to improve his apparatus. He was incessantly making fresh mirrors, or trying new lenses, or combinations of lenses to act as eye-pieces, or projecting alterations in the mounting by which the telescope was supported. Such was his enthusiasm that his house, we are told, was incessantly littered with the usual indications of the workman's presence, greatly to the distress of his sister, who, at this time, had come to take up her abode with him and look after his housekeeping. Indeed, she complained that in his astronomical ardour he sometimes omitted to take off, before going into his workshop, the beautiful lace ruffles which he wore while conducting a concert, and that consequently they became soiled with the pitch employed in the polishing of his mirrors.

- ROBERT STAWELL BALL

This sister, who occupies such a distinct place in scientific history, is the same little girl to whom we have already referred. From her earliest days she seems to have cherished a passionate admiration for her brilliant brother William. It was the proudest delight of her childhood as well as of her mature years to render him , whatever service she could; no man of science was ever provided with a more capable or energetic helper than William Herschel found in this remarkable woman. Whatever work had to be done she was willing to bear her share in it, or even to toil at it unassisted if she could be allowed to do so. She not only managed all his domestic affairs, but in the grinding of the lenses and in the polishing of the mirrors she rendered every assistance that was possible. At one stage of the very delicate operation of fashioning a reflector, it is necessary for the workman to remain with his hand on the mirror for many hours in succession. When such labours were in progress, Caroline used to sit by her brother, and enliven the time by reading stories aloud, sometimes pausing to feed him with a spoon while his hands were engaged on the task from which he could not desist for a moment.

When mathematical work had to be done Caroline was ready for it; she had taught herself sufficient to enable her to perform the kind of calculations, not, perhaps, very difficult ones, that Herschel's work required; indeed, it is not too much to say that the mighty life-work which this man was enabled to perform could never have been accomplished had it not been for the self-sacrifice of this ever-loving and faithful sister. When Herschel was at the telescope at night, Caroline sat by him at her desk, pen in hand, ready to write down the notes of the observations as they fell from her brother's lips. This was no insignificant toil. The telescope was, of course, in the open air, and as Herschel not unfrequently continued his observations throughout the whole of a long winter's night, there were but few women who could have accomplished the task which Caroline so cheerfully executed. From dusk till dawn, when the sky was clear, were Herschel's observing hours, and what this sometimes implied we can realise from the fact that Caroline assures us she had sometimes to desist because the ink had actually frozen in her pen. The night's work over, a brief rest was taken, and while William had his labours for the day to attend to, Caroline carefully transcribed the observations made during the night before, reduced all the figures, and prepared everything in readiness for the observations that were to follow on the ensuing evening.

But we have here been anticipating a little of the future which lay before the great astronomer; we must now revert to the history of his early work, at Bath, in 1774, when Herschel's scrutiny of the skies first commenced with an instrument of his own manufacture. For some few years he did not attain any result of importance; no doubt he made a few interesting observations, but the value of the work during those years is to be found, not in any actual discoveries which were accomplished, but in the practice which Herschel obtained in the use of his instruments. It was not until 1782 that the great achievement took place by which he at once sprang into fame.

Garden view, Herschel House, Slough.

It is sometimes said that discoveries are made by accident, and, no doubt, to a certain extent, but only, I fancy, to a very small extent, this statement may be true. It is, at all events, certain that such lucky accidents do not often fall to the lot of people unless those people have done much to deserve them. This was certainly the case with Herschel. He appears to have formed a project for making a close examination of all the stars above a certain magnitude. Perhaps he intended to confine this research to a limited region of the sky, but, at all events, he seems to have undertaken the work energetically and systematically. Star after star was brought to the centre of the field of view of his telescope, and after being carefully examined was then displaced, while another star was brought forward to be submitted to the same process. In the great majority of cases such observations yield really nothing of importance; no doubt even the smallest star in the heavens would, if we could find out all about it, reveal far more than all the astronomers that were ever on the earth have even conjectured. What we actually learn about the great majority of stars is only information of the most meagre description. We see that the star is a little point of light, and we see nothing more.

In the great review which Herschel undertook he doubtless examined hundreds, or perhaps thousands of stars, allowing them to pass away without note or comment. But on an ever-memorable night in March, 1782, it happened that he was pursuing his task among the stars in the Constellation of Gemini. Doubtless, on that night, as on so many other nights, one star after another was looked at only to be dismissed, as not requiring further attention. On the evening in question, however, one star was noticed which, to Herschel's acute vision, seemed different from the stars which in so many thousands are strewn over the sky. A star properly so called appears merely as a little point of light, which no increase of magnifying power will ever exhibit with a true disc. But there was something in the star-like object which Herschel saw that immediately arrested his attention and made him apply to it a higher magnifying power. This at once disclosed the fact that the object possessed a disc, that is, a definite, measurable size, and that it was thus totally different from any one of the hundreds and thousands of stars which exist elsewhere in space. Indeed, we may say at once that this little object was not a star at all; it was a planet. That such was its true nature was confirmed, after a little further observation, by perceiving that the body was shifting its place on the heavens relatively to the stars. The organist at the Octagon Chapel at Bath had, therefore, discovered a new planet with his home-made telescope.

I can imagine someone will say, "Oh, there was nothing so wonderful in that; are not planets always being discovered? Has not

M. Palisa, for instance, discovered about eighty of such objects, and are there not hundreds of them known nowadays?" This is, to a certain extent, quite true. I have not the least desire to detract from the credit of those industrious and sharp-sighted astronomers who have in modern days brought so many of these little objects within our cognisance. I think, however, it must be admitted that such discoveries have a totally different importance in the history of science from that which belongs to the peerless achievement of Herschel. In the first place, it must be observed that the minor planets now brought to light are so minute that if a score of them were rolled together into one lump it would not be one-thousandth part of the size of the grand planet discovered by Herschel. This is, nevertheless, not the most important point. What marks Herschel's achievement as one of the great epochs in the history of astronomy is the fact that the detection of Uranus was the very first recorded occasion of the discovery of any planet whatever.

For uncounted ages those who watched the skies had been aware of the existence of the five old planets Jupiter, Mercury, Saturn, Venus, and Mars. It never seems to have occurred to any of the ancient philosophers that there could be other similar objects as yet undetected over and above the well-known five. Great then was the astonishment of the scientific world when the Bath organist announced his discovery that the five planets which had been known from all antiquity must now admit the company of a sixth. And this sixth planet was, indeed, worthy on every ground to be received into the ranks of the five glorious bodies of antiquity. It was, no doubt, not so large as Saturn, it was certainly very much less than Jupiter; on the other hand, the new body was very much larger than Mercury, than Venus, or than Mars, and the earth itself seemed quite an insignificant object in comparison with this newly added member of the Solar

System. In one respect, too, Herschel's new planet was a much more imposing object than any one of the older bodies; it swept around the sun in a majestic orbit, far outside that of Saturn, which had previously been regarded as the boundary of the Solar System, and its stately progress required a period of not less than eighty-one years.

King George the Third, hearing of the achievements of the Hanoverian musician, felt much interest in his discovery, and accordingly Herschel was bidden to come to Windsor, and to bring with him the famous telescope, in order to exhibit the new planet to the King, and to tell his Majesty all about it. The result of the interview was to give Herschel the opportunity for which he had so long wished, of being able to devote himself exclusively to science for the rest of his life.

- ROBERT STAWELL BALL

View of the Observatory, Herschel House, Slough.

The King took so great a fancy to the astronomer that he first, as I have already mentioned, duly pardoned his desertion from the army, some twenty-five years previously. As a further mark of his favour the King proposed to confer on Herschel the title of his Majesty's own astronomer, to assign to him a residence near Windsor, to provide him with a salary, and to furnish such funds as might be required for the erection of great telescopes, and for the conduct of that mighty scheme of celestial observation on which Herschel was so eager to enter. Herschel's capacity for work would have been much impaired if he had been deprived of the aid of his admirable sister, and to her, therefore, the King also assigned a salary, and she was installed as Herschel's assistant in his new post.

With his usually impulsive determination, Herschel immediately cut himself free from all his musical avocations at Bath, and at once entered on the task of making and erecting the great telescopes at Windsor. There, for more than thirty years, he and his faithful sister prosecuted with unremitting ardour their nightly scrutiny of the sky. Paper after paper was sent to the Royal Society, describing the hundreds, indeed the thousands, of objects such as double stars; nebulse and clusters, which were first revealed to human gaze during those midnight vigils. To the end of his life he still continued at every possible opportunity to devote himself to that beloved pursuit in which he had such unparalleled success. No single discovery of Herschel's later years was, however, of the same momentous description as that which first brought him to fame.

The 40-foot telescope as it was in the year 1863. Herschel House, Slough.

Herschel married when considerably advanced in life, and he lived to enjoy the indescribable pleasure of finding that his only son, afterwards Sir John Herschel, was treading worthily in his footsteps, and attaining renown as an astronomical observer, second only to that of his father. The elder Herschel died in 1822, and his illustrious sister Caroline then returned to Hanover, where she lived for many years to receive the respect and attention which were so justly hers. She died at a very advanced age in 1848.

- ROBERT STAWELL BALL

COMPLIMENTARY MATERIAL

A BRIEF BIOGRAPHICAL SKETCH

WILLIAM HERSCHEL

- ROBERT STAWELL BALL

William Herschel
(November 15, 1738 – August 25, 1822)

COMPLIMENTARY MATERIAL

A BRIEF BIOGRAPHICAL SKETCH

WILLIAM HERSCHEL

Frederick William Herschel, KH, FRS (November 15, 1738 – August 25, 1822) was a British astronomer and composer of German origin, and brother of fellow astronomer Caroline Herschel, with whom he worked. Born in the Electorate of Hanover, Herschel followed his father into the Military Band of Hanover, before migrating to Great Britain in 1757 at the age of nineteen.

Herschel constructed his first large telescope in 1774, after which he spent nine years carrying out sky surveys to investigate double stars. The resolving power of the Herschel telescopes revealed that the nebulae in the Messier catalogue were clusters of stars. Herschel published catalogues of nebulae in 1802 (2,500 objects) and in 1820 (5,000 objects). In the course of an observation on 13 March 1781, he realized that one celestial body he had observed was not a star, but a planet, Uranus. This was the first planet to be discovered since antiquity and Herschel became famous overnight. As a result of this discovery, George III appointed him Court Astronomer. He was elected as a Fellow of the Royal Society and grants were provided for the construction of new telescopes.

Herschel pioneered the use of astronomical spectrophotometry as a diagnostic tool, using prisms and temperature measuring equipment to measure the wavelength distribution of stellar spectra. Other work included an improved determination of the rotation period of Mars, the discovery that the Martian polar caps vary seasonally, the discovery of Titania and Oberon (moons of Uranus) and Enceladus and Mimas (moons of Saturn). In addition, Herschel discovered infrared radiation. Herschel was made a Knight of the Royal Guelphic Order in 1816. He was the first

President of the Royal Astronomical Society when it was founded in 1820. He died in August 1822, and his work was continued by his only son, John Herschel.

Early Life and Musical Activities

Herschel was born in the Electorate of Hanover in Germany, then part of the Holy Roman Empire, one of ten children of Isaac Herschel by his marriage to Anna Ilse Moritzen. His family were Lutheran Christians. His father was an oboist in the Hanover Military Band. In 1755 the Hanoverian Guards regiment, in whose band Wilhelm and his brother Jakob were engaged as oboists, was ordered to England. At the time the crowns of Great Britain and Hanover were united under King George II. As the threat of war with France loomed, the Hanoverian Guards were recalled from England to defend Hanover. After they were defeated at the Battle of Hastenbeck, Herschel's father Isaak sent his two sons to seek refuge in England in late 1757. Although his older brother Jakob had received his dismissal from the Hanoverian Guards, Wilhelm was accused of desertion (for which he was pardoned by George III in 1782). Wilhelm, nineteen years old at this time, was a quick student of the English language. In England he went by the English rendition of his name, Frederick William Herschel.

In addition to the oboe, he played the violin and harpsichord and later the organ. He composed numerous musical works, including 24 symphonies and many concertos, as well as some church music. Six of his symphonies were recorded in April 2002 by the London Mozart Players, conducted by Matthias Bamert (Chandos 10048).

Herschel moved to Sunderland in 1761 when Charles Avison immediately engaged him as first violin and soloist for his Newcastle orchestra, where he played for one season. In 'Sunderland in the County of Durh: apprill [sic] 20th 1761' he wrote his Symphony No. 8 in C Minor. He was head of the Durham Militia band 1760–61 and visited the home of Sir Ralph Milbanke at Halnaby Hall in 1760, where he wrote two symphonies, as well as giving performances himself.

Original manuscript of Symphony No. 15 in E flat major (1762).

After Newcastle he moved to Leeds and Halifax where he was the first organist at St John the Baptist church (now Halifax Minster). He became organist of the Octagon Chapel, Bath, a fashionable chapel in a well-known spa, in which city he was also Director of Public Concerts. He was appointed as the organist in 1766 and gave his introductory concert on 1 January 1767. As the organ was still incomplete, he showed off his versatility by performing his own compositions including a violin concerto, an oboe concerto and a harpsichord sonata. The organ was completed in October 1767. His sister Caroline came to England in 1772 and lived with him there in New King Street, Bath. The house they shared is now the location of the Herschel Museum of Astronomy. His brothers Dietrich, Alexander and Jakob (1734–1792) also appeared as musicians of Bath. In 1780, Herschel was appointed director of the Bath orchestra, with his sister often appearing as soprano soloist.

Herschel's music led him to make the acquaintance of an amateur violinist, the Rev. John Michell of Thornhill, a former leading Cambridge mathematics professor who was already noted as England's leading geologist and was developing groundbreaking views on astronomy and the construction of telescopes. Michell sparked Herschel's interest in mathematics and lenses. Michell was later (1783) to publish work on 'dark stars' (black holes). After Michell's death in 1793, Herschel bought a ten foot long, 30-inch reflecting telescope from Michell's estate which he had to reconstruct due to damage in transport.

His interest in astronomy grew stronger after he made the acquaintance of the English Astronomer Royal Nevil Maskelyne. He started building his own reflecting telescopes and would spend up to 16 hours a day grinding and polishing the speculum metal primary mirrors. He "began to look at the planets and the stars" in May 1773 and on 1 March 1774 began an astronomical journal by noting his observations of Saturn's rings and the Great Orion Nebula (M 42).

Replica in the **William Herschel Museum**, Bath, of a telescope similar to that with which Herschel discovered **Uranus**

Double Stars

Herschel's early observational work soon focused on the search for pairs of stars that were very close together visually. Astronomers of the era expected that changes over time in the apparent separation and relative location of these stars would provide evidence for both the proper motion of stars and, by means of parallax shifts in their separation, for the distance of stars from the Earth. The latter was a method first suggested by Galileo Galilei. From the back garden of his house in New King Street, Bath, and using a 6.2-inch aperture (160 mm), 7-foot focal length (2.1 m) (f/13) Newtonian telescope "with a most capital speculum" of his own manufacture, in October 1779, Herschel began a systematic search for such stars among "every star in the Heavens", with new discoveries listed through 1792. He soon discovered many more binary and multiple stars than expected, and compiled them with careful measurements of their relative positions in two catalogues presented to the Royal Society in London in 1782 (269 double or multiple systems) and 1784 (434 systems). A third catalogue of discoveries made after 1783 was published in 1821 (145 systems).

In 1797, Herschel measured many of the systems again, and discovered changes in their relative positions that could not be attributed to the parallax caused by the Earth's orbit. He waited until 1802 (in Catalogue of 500 new Nebulae, nebulous Stars, planetary Nebulae, and Clusters of Stars; with Remarks on the Construction of the Heavens) to announce the hypothesis that the two stars might be "binary sidereal systems" orbiting under mutual gravitational attraction, a hypothesis he confirmed in 1803 in his Account of the Changes that have happened, during the last Twenty-five Years, in the relative Situation of Double-stars; with an Investigation of the Cause to which they are owing. In all, Herschel discovered over 800 confirmed double or multiple star systems, almost all of them physical rather than optical pairs. His theoretical and observational work provided the foundation for modern binary star astronomy; new catalogues adding to his work were not published until after 1820 by Friedrich Wilhelm Struve, James South and John Herschel.

Uranus

In March 1781, during his search for double stars, Herschel noticed an object appearing as a disk. Herschel originally thought it was a comet or a stellar disk, which he believed he might actually resolve.[14] He made many more observations of it, and afterwards Russian Academician Anders Lexell computed the orbit and found it to be probably planetary. Herschel agreed, determining that it must be a planet beyond the orbit of Saturn. He called the new planet the 'Georgian star' (Georgium sidus) after King George III, which also brought him favour; the name did not stick. In France, where reference to the British king was to be avoided if possible, the planet was known as 'Herschel' until the name 'Uranus' was universally adopted. The same year, Herschel was awarded the Copley Medal and elected a Fellow of the Royal Society.

In 1782, he was appointed "The King's Astronomer" (not to be confused with the Astronomer Royal). He and his sister subsequently moved to Datchet (then in Buckinghamshire but now in Berkshire) on 1 August 1782. He continued his work as a telescope maker and achieved an international reputation for their manufacture, profitably selling over 60 completed reflectors to British and Continental astronomers.

Deep Sky Surveys

From 1782 to 1802, and most intensively from 1783 to 1790, Herschel conducted systematic surveys in search of "deep sky" or nonstellar objects with two 20-foot focal length (610 cm), 12-inch aperture (30 cm) and 18.7-inch aperture (47 cm) telescopes (in combination with his favoured 6-inch aperture instrument). Excluding duplicated and "lost" entries, Herschel ultimately discovered over 2400 objects defined by him as nebulae. (At that time, nebula was the generic term for any visually diffuse astronomical object, including galaxies beyond the Milky Way, until galaxies were confirmed as extragalactic systems by Edwin Hubble in 1924.)

NGC 2683 is an unbarred spiral galaxy discovered by William Herschel on 5 February 1788.

Herschel published his discoveries as three catalogues: Catalogue of One Thousand New Nebulae and Clusters of Stars (1786), Catalogue of a Second Thousand New Nebulae and Clusters of Stars (1789) and the previously cited Catalogue of 500 New

Nebulae ... (1802). He arranged his discoveries under eight "classes": (I) bright nebulae, (II) faint nebulae, (III) very faint nebulae, (IV) planetary nebulae, (V) very large nebulae, (VI) very compressed and rich clusters of stars, (VII) compressed clusters of small and large [faint and bright] stars, and (VIII) coarsely scattered clusters of stars. Herschel's discoveries were supplemented by those of Caroline Herschel (11 objects) and his son John Herschel (1754 objects) and published by him as General Catalogue of Nebulae and Clusters in 1864. This catalogue was later edited by John Dreyer, supplemented with discoveries by many other 19th century astronomers, and published in 1888 as the New General Catalogue (abbreviated NGC) of 7840 deep sky objects. The NGC numbering is still the most commonly used identifying label for these celestial landmarks.

Work with his sister Caroline

William and Caroline William Herschel polishing a telescope lens (probably a mirror), 1896 Lithograph.

Following the death of their father, William suggested that Caroline join him in Bath, England. In 1772, Caroline was first introduced to astronomy by her brother.

Caroline spent many hours polishing the mirrors of high performance telescopes so that the amount of light captured was maximized. She also copied astronomical catalogues and other publications for William. After William accepted the office of King's Astronomer to George III, Caroline became his constant assistant.

In October 1783, a new 20-foot telescope came into service for William. During this time, William was attempting to observe and then record all of the observations. He had to run inside and let his eyes readjust to the artificial light before he could record anything, and then he would have to wait until his eyes were adjusted to the dark before he could observe again. Caroline became his recorder by sitting at a desk near an open window. William would shout out his observations and she would write them down along with any information he needed from a reference book.

In 1783, he gave Caroline a telescope, and she began to make astronomical discoveries in her own right, particularly comets. Between 1783 and 1787, she made an independent discovery of M110 (NGC 205), which is the second companion of the Andromeda Galaxy. During the years 1786-1797, she discovered or observed eight comets, eleven nebulae and, at her brother's suggestion, updated and corrected Flamsteed's work detailing the position of stars. Her eight comets were published between 28 August 1782 to 5 February 1787. Five of her comets were published in Philosophical Transactions of the Royal Society. William was even summoned to Windsor Castle to demonstrate Caroline's comet to the royal family. William recorded this phenomenon himself, terming it "My Sister's Comet." She wrote letters to the Astronomer Royal to announce the discovery of her second comet, and wrote to Joseph Banks upon the discovery of her third and fourth comets. Flamsteed's work was published as the British Catalogue of Stars. The catalogue contained an index of more than 560 stars that had not been previously included. She also rediscovered Comet Encke in 1795. She was honoured by the Royal Astronomical Society for this work in 1828. Caroline also continued to serve as his assistant, often taking notes while he observed at the telescope.

For her work as William's assistant, she was granted an annual salary of £50 (equivalent to £5,700 in 2016) by George III. Her appointment made her the first female in England to be honored with a government position.

William and Caroline William Herschel
polishing a telescope lens (probably a mirror),
1896 Lithograph.

In June 1785, owing to damp conditions, he and Caroline moved to Clay Hall in Old Windsor. In 1786, the Herschels moved to a new residence on Windsor Road in Slough. He lived the rest of his life in this residence, which came to be known as Observatory House. It is no longer standing.

On 8 May 1788, he married the widow Mary Pitt (née Baldwin) at St Laurence's Church, Upton in Slough. The marriage caused a lot of tension in the brother-sister relationship. Caroline has been referred to as a bitter, jealous woman who worshiped her brother and resented her sister-in-law for invading her domestic life. With the arrival of Mary, Caroline lost her managerial and social responsibilities in the household. Caroline destroyed her journals between the years 1788 to 1798, so her feelings during this period are not entirely known. According to her memoir, Caroline then moved to separate lodgings, but continued to work as her brother's assistant. When her brother and his family were away from their home, she would often return to take care of it for them. In later life, Caroline and Lady Herschel exchanged affectionate letters.

Caroline continued to work on her astronomical work, even after William's death in 1822. She worked to verify and confirm his findings as well as putting together catalogues of nebulae. Towards the end of her life, she arranged two-and-a-half thousand nebulae and star clusters into zones of similar polar distances. She did this so that her nephew, John, could re-examine them systematically. Eventually, this list was enlarged and renamed the New General Catalogue. In 1828, she was awarded the Royal Astronomical Society's Gold Medal for her work.

Herschel's Telescopes

During his career, he constructed more than four hundred telescopes. The largest and most famous of these was a reflecting telescope with a 49 1⁄2-inch-diameter (1.26 m) primary mirror and a 40-foot (12 m) focal length. Because of the poor

reflectivity of the speculum mirrors of that day, Herschel eliminated the small diagonal mirror of a standard newtonian reflector from his design and tilted his primary mirror so he could view the formed image directly. This design has come to be called the Herschelian telescope. In 1789, shortly after this instrument was operational, he discovered a new moon of Saturn: Mimas, only 250 miles in diameter. A second moon followed within the first month of observation. The "40-foot telescope" proved very cumbersome, and most of his observations were done with a smaller 18.5-inch (47 cm) 20-foot-focal-length (6.1 m) reflector. Herschel discovered that unfilled telescope apertures can be used to obtain high angular resolution, something which became the essential basis for interferometric imaging in astronomy (in particular Aperture Masking Interferometry and hypertelescopes).
Reconstruction of the 20ft telescope

In 2012, the BBC Stargazing Live television programme built a replica of the 20-foot telescope using Herschel's original plans but modern materials. It is to be considered a close modern approximation rather than an exact replica. A modern glass mirror was used, the frame uses metal scaffolding and the tube is a sewer pipe. The telescope was shown on the programme in January 2013 and stands on the Art, Design and Technology campus of the University of Derby where it will be used for educational purposes.

Life on other Celestial Bodies

Herschel was sure that he had found ample evidence of life on the Moon and compared it to the English countryside. He did not refrain himself from theorising that the other planets were populated, with a special interest in Mars, which was in line with most of his contemporary scientists. At Herschel's time, scientists tended to believe in a plurality of civilised worlds; in contrast, most religious thinkers referred to unique properties of the earth. Herschel went so far to speculate that the interior of the sun was populated.

Sunspots, Climate, and Wheat Yields

Herschel started to examine the correlation of solar variation and solar cycle and climate. Over a period of 40 years (1779–1818), Herschel had regularly observed sunspots and their variations in number, form and size. Most of his observations took place in a period of low solar activity, the Dalton minimum, when sunspots were relatively few in number. This was one of the reasons why Herschel was not able to identify the standard 11-year period in solar activity. Herschel compared his observations with the series of wheat prices published by Adam Smith in The Wealth of Nations.

In 1801, Herschel reported his findings to the Royal Society and indicated five prolonged periods of few sunspots correlated with the price of wheat. Herschel's study was ridiculed by some of his contemporaries but did initiate further attempts to find a correlation. Later in the 19th century, William Stanley Jevons proposed the 11-year cycle with Herschel's basic idea of a correlation between the low amount of sunspots and lower yields explaining recurring booms and slumps in the economy. Herschel's speculation on a connection between sunspots and regional climate, using the market price of wheat as a proxy, continues to be cited.

According to one study, the influence of solar activity can actually be seen on the historical wheat market in England over ten solar cycles between 1600 and 1700. The evaluation is controversial and the significance of the correlation is doubted by some scientists.

Further Discoveries

Moons discovered: 4

Oberon	11 January 1787
Titania	11 January 1787
Enceladus	28 August 1789
Mimas	17 September 1789

Planets discovered: 1

Uranus	13 March 1781

In his later career, Herschel discovered two moons of Saturn, Mimas and Enceladus; as well as two moons of Uranus, Titania and Oberon. He did not give these moons their names; they were named by his son John in 1847 and 1852, respectively, after his death. Herschel measured the axial tilt of Mars and discovered that the martian ice caps, first observed by Giovanni Domenico Cassini (1666) and Christiaan Huygens (1672), changed size with that planet's seasons. It has been suggested that Herschel might have discovered rings around Uranus.

Herschel introduced but did not create the word "asteroid", meaning star-like (from the Greek asteroeides, aster "star" + -eidos "form, shape"), in 1802 (shortly after Olbers discovered the second minor planet, 2 Pallas, in late March), to describe the star-like appearance of the small moons of the giant planets and of the minor planets; the planets all show discs, by comparison. By the 1850s 'asteroid' became a standard term for describing certain minor planets.

From studying the proper motion of stars, Herschel was the first to realize that the solar system is moving through space, and he determined the approximate direction of that movement. He also studied the structure of the Milky Way and concluded that it was in the shape of a disk. He incorrectly assumed the sun was in the centre of the disc, a theory known as Galactocentrism, which was eventually corrected by the findings of Harlow Shapley in 1918.

Discovery of Infrared Radiation in Sunlight

On 11 February 1800, Herschel was testing filters for the sun so he could observe sun spots. When using a red filter he found there was a lot of heat produced. Herschel discovered infrared radiation in sunlight by passing it through a prism and holding a thermometer just beyond the red end of the visible spectrum. This thermometer was meant to be a control to measure the ambient air temperature in the room. He was shocked when it showed a higher temperature than the visible spectrum. Further experimentation led to Herschel's conclusion that there must be an invisible form of light beyond the visible spectrum.

Biology

Herschel used a microscope to establish that coral was not a plant--as many at the time believed--because it lacked the cell walls characteristic of plants.

Family and Death

William Herschel and Mary had one child, John, born at Observatory House on 7 March 1792. William's personal background and rise as man of science had a profound impact on the upbringing of his son and grandchildren. He was elected a Foreign Honorary Member of the American Academy of Arts and Sciences in 1788. In 1816, William was made a Knight of the Royal Guelphic Order by the Prince Regent and was accorded the honorary title 'Sir' although this was not the equivalent of an official British knighthood. He helped to found the Astronomical Society of London in 1820, which in 1831 received a royal charter and became the Royal Astronomical Society. In

1813, he was elected a foreign member of the Royal Swedish Academy of Sciences.

On 25 August 1822, Herschel died at Observatory House, Windsor Road, Slough, and was buried at nearby St Laurence's Church, Upton, Slough. Herschel's epitaph is

> " *Coelorum perrupit claustra*
> *(He broke through the barriers*
> *of the heavens)* "

Herschel's son John Herschel also became a famous astronomer. One of William's brothers, Alexander Herschel, moved permanently to England, near his sister Caroline and nephew John. Caroline returned to Hanover after the death of her brother. She died on 9 January 1848.

His house at 19 New King Street in Bath, Somerset, where he made many telescopes and first observed Uranus, is now home to the Herschel Museum of Astronomy.

Memorial

William Herschel lived most of his life in Slough, a town then in Buckinghamshire. He died in the town and was buried under the tower of the Church of St Laurence, Upton-cum-Chalvey, near Slough. Herschel is quite respected in the town and there are several memorials to him and his discoveries. In 2011 a new bus station, the design of which was inspired by the infrared experiment of William Herschel, was built in the centre of Slough.

Musical Works

Herschel's complete musical works were as follows:

- 18 symphonies for small orchestra (1760–1762)
- 6 symphonies for large orchestra (1762–1764)

- 12 concertos for oboe, violin and viola (1759–1764)

- 2 concertos for organ

- 6 sonatas for violin, cello and harpsichord (published 1769)

- 12 solos for violin and basso continuo (1763)

- 24 capriccios and 1 sonata for solo violin

- 1 andante for two basset horns, two oboes, two horns and two bassoons.

- Various vocal works including a "Te Deum", psalms, motets and sacred chants along with some catches.

- Keyboard works for organ and harpsichord:

- 6 fugues for organ

- 24 sonatas for organ (10 now lost)

- 33 voluntaries and pieces for organ (incomplete)
- 24 pieces for organ (incomplete)
- 12 voluntaries (11 now lost)
- 12 sonatas for harpsichord (9 extant)
- 25 variations on an ascending scale
- 2 minuets for harpsichord

Named after Herschel

- William Herschel, portrait by James Sharples, c. 1805
- The astrological symbol for planet Uranus (Uranus's astrological symbol.svg) features the capital initial letter of Herschel's surname.
- Mu Cephei is also known as Herschel's Garnet Star
- Herschel, a crater on the Moon

- Herschel, a large impact basin on Mars
- The enormous crater Herschel on Saturn's moon Mimas
- 2000 Herschel, an asteroid
- The William Herschel Telescope on La Palma
- The Herschel Space Observatory, successfully launched by the European Space Agency on 14 May 2009. It is the largest space telescope of its kind
- Herschel Grammar School, Slough
- Rue Herschel, a street in the 6th Arrondissement of Paris.
- The Herschel Building at Bath College, Bath
- The Herschel building at Newcastle

University, Newcastle, United Kingdom

- Herschel Museum of Astronomy, at 19 New King Street in Bath.

- Herschelschule, Hanover, Germany, a grammar school

- Herschel, Saskatchewan, Canada is a small, rural village that is home to the Ancient Echoes Interpretative Centre

- The Herschel Observatory, at the Universitas School in Santos, Brazil.

- The lunar crater C. Herschel, the asteroid 281 Lucretia, and the comet 35P/Herschel-Rigollet are named after his sister Caroline Herschel.

- The public house "Herschel Arms" at 22 Park Street, Slough is named after him and is quite close to the site of Observatory House.

- Herschel Astronomical Society, the

operator of the Herschel Memorial Observatory based in Eton, Berkshire.

- Herschel Park, Slough.

- The shape of Slough Bus Station, built in 2011, was inspired by Herschel's infrared experiment.

ABOUT THE AUTHOR

ROBERT STAWELL BALL
(July 1, 1840 – November 25, 1913)

Sir Robert Stawell Ball was an Irish astronomer who founded the screw theory.

He was the son of naturalist Robert Ball and Amelia Gresley Hellicar. He was born in Dublin.

Ball worked for Lord Rosse from 1865 to 1867. In 1867 he became Professor of Applied Mathematics at the Royal College of Science in Dublin. There he lectured on mechanics and published an elementary account of the science.

In 1874 Ball was appointed Royal Astronomer of Ireland and Andrews Professor of Astronomy in the University of Dublin at Dunsink Observatory.

Ball contributed to the science of kinematics by delineating the screw displacement:

When Ball and the screw theorists speak of screws they no longer mean actual cylindrical objects with helical threads cut into them but the possible motion of any body whatsoever, including that of the screw independently of the nut.

Ball's treatise *The Theory of Screws* (1876) is now in the public domain. His work on screw dynamics earned him in 1879 the

Cunningham Medal of the Royal Irish Academy.

In 1882 Popular Science Monthly carried his article "*A Glimpse through the Corridors of Time*". The following year it carried his two-part article on "*The Boundaries of Astronomy*".

Ball expounded the tides in *Time and Tide*: a Romance of the Moon In 1892 he was appointed Lowndean Professor of Astronomy and Geometry at Cambridge University at the same time becoming director of the Cambridge Observatory. He was a fellow of King's College, Cambridge.

In 1900 Cambridge University Press published *A Treatise on the Theory of Screws*. That year he also published *The Story of the Heavens*. Much in the limelight, he stood as President of the Quaternion Society. He was also President of the Mathematical Association in 1900.

In 1908 he published *A Treatise on Spherical Astronomy*, which is a textbook on astronomy starting from spherical trigonometry and the celestial sphere, considering atmospheric refraction and aberration of light, and introducing basic use of a generalised instrument.

His work *The Story of the Heavens* is mentioned in the "Ithaka" chapter of Ulysses. His lectures, articles and books (e.g. *Starland* and *The Story of the Heavens*) were mostly popular and simple in style.

He died in Cambridge and was buried at the Parish of the Ascension Burial Ground in Cambridge, with his wife Lady Francis Elizabeth Ball. Their children were: Frances Amelia, Robert Steele, William Valentine (later Sir), Mary Agnetta, Charles Rowan Hamilton, and Randall Gresley (later Colonel).

LECTURES

Ball became celebrated for his popular lectures on science. He gave an estimated 2500 lectures between 1875 and 1910 in towns and cities across Britain and Ireland. In 1892, 1898 and 1900 he was invited to deliver the Royal Institution Christmas Lecture. *Astronomy; Astronomy* and *Great Chapters from the Book of Nature.*

FEEDBACK

Now that you have read the book ...

Was it interesting?

Did you enjoy what you wanted to read?
Was there any room for improvement?

Let us know at:

http://www.diamondbooks.ca/feedback

Your feedback is highly appreciated.
Thank you!

NOTES

NOTES

NOTES

NOTES

Printed in Great Britain
by Amazon

41622268R00066